T0187269

2nd EDITION

Activity Book
Foundation A

Trace and join

Date: _____

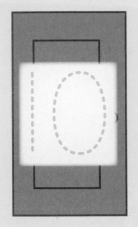

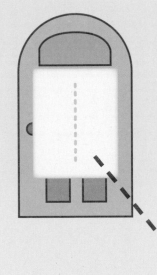

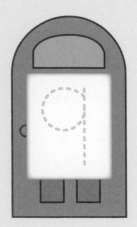

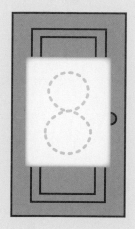

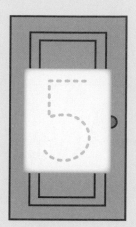

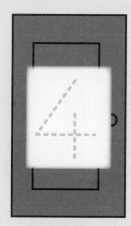

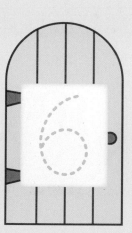

Teacher's notes

Trace the number on each door. Then draw lines to join the doors in order, from 1 to 10.

Date: _____

Count and match

2

1

5

4

3

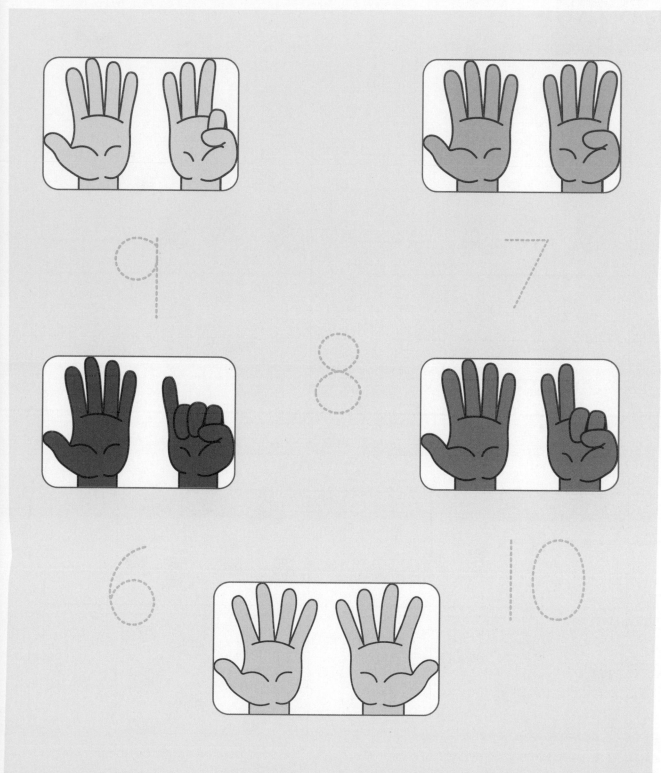

Teacher's notes

Count the fingers on each hand. Draw a line to the matching number. Then trace the number.

Date: _____

Fewer

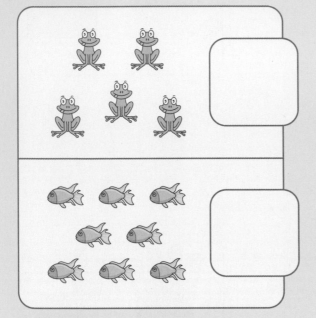

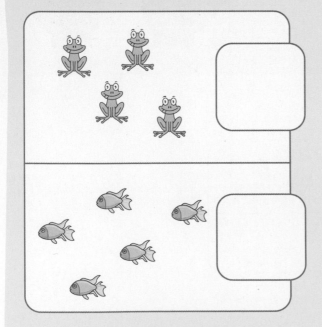

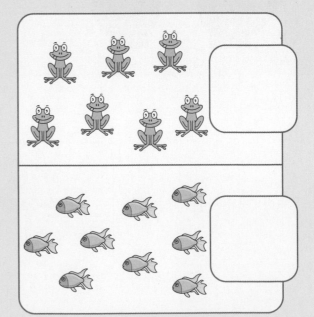

Teacher's notes

For each box, count the frogs and fish. Put a tick next to the box that has **fewer** creatures.

More

Date: _____

| 1 | 2 | 3 | 4 | 5 | 6 | 7 | 8 | 9 | 10 |

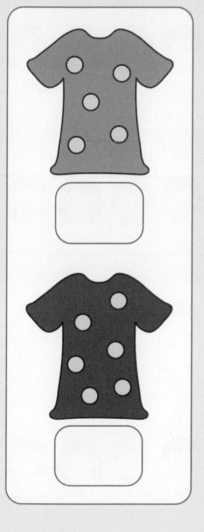

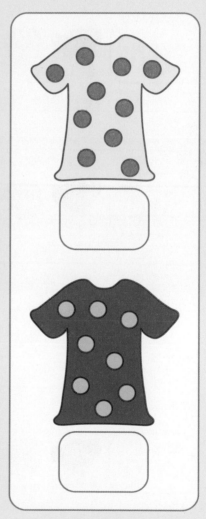

Teacher's notes

Count the spots on each T-shirt. Write the number in the box. For each pair of T-shirts, circle the number that is **more**.

7

Ordinals

Date: _____

1st	2nd	3rd	4th	5th	6th	7th	8th	9th	10th

8th

1st

3rd

5th

2nd

Teacher's notes

Look at the ordinal number on the card in each row. Circle the child in that position.

More

Date: _____

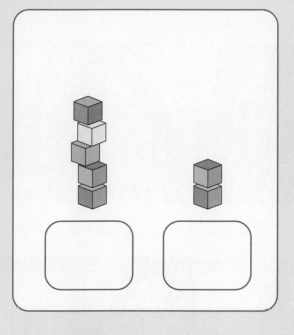

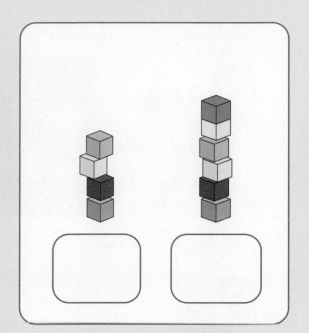

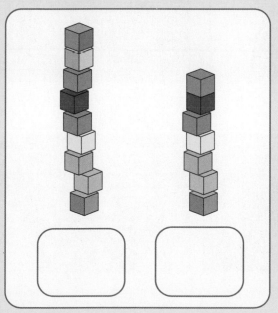

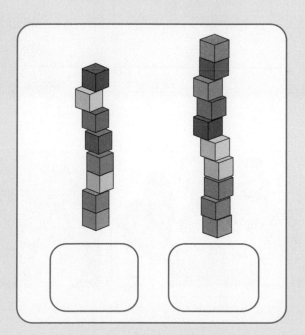

Teacher's notes

Count the blocks in each tower. Write the number in the box. In each pair, circle the tower with **more** blocks.

9

Fewest and most

Date: _____

Date: _____

Smallest and largest

0 1 2 3 4 5 6 7 8 9 10

0 1 3

7 9 6

8 2 4

5 10 7

6 3 0

8 4 9

11

Position

Date: _____

Teacher's notes

Colour **red** the fish in the **middle** of the tank. Colour **yellow** a fish in a **corner**. Colour **purple** a fish at an **edge**. Draw a fish **between** two fish you have coloured.

Date: _____

Direction

1	2	3	4	5

1	2	3	4	5

5
4
3
2
1

5
4
3
2
1

Teacher's notes

The rabbit jumps **backwards** 2 spaces. The frog jumps **forwards** 3 spaces. The bird flies **up** 2 spaces. The bee flies **down** 1 space. Circle the number each creature lands on.

13

Left and right

Date: _____

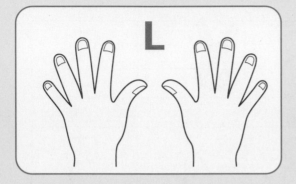

L

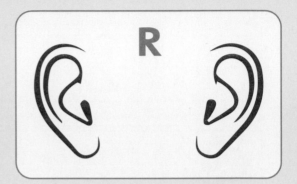

R

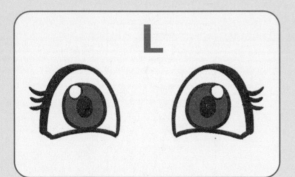

L

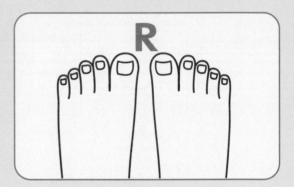

R

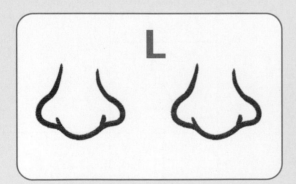

L

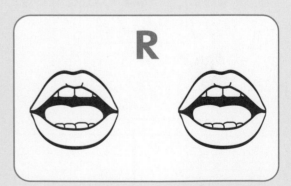

R

Teacher's notes

For the boxes on the left, colour the left-hand object. For the boxes on the right, colour the right-hand object.

14

Following instructions

Date: _____

left 2

up 1

down 2

right 3

Teacher's notes

Begin at the flower for each separate instruction. Circle the creature you land on.

15

Longest

Date: _____

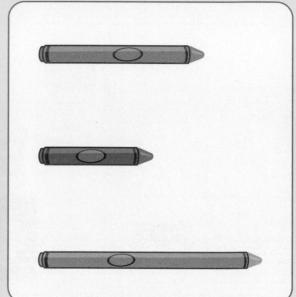

Date: _____

Widest

  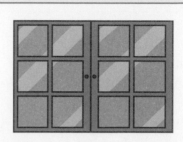

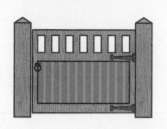 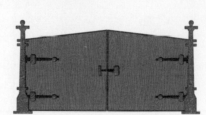 

Date: _____

Tallest

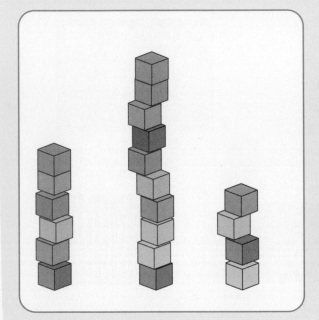

Teacher's notes

Circle the tallest object or animal in each set.

Highest

Date: _____

Teacher's notes

Circle the highest creature in each picture.

Sort 2D shapes

Date: _____

Teacher's notes

Colour **green** all the shapes with **1 side**. Colour **orange** all the shapes with **3 sides**. Colour **purple** all the shapes with **4 sides**.

Pyramids and cones

Date: _____

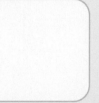

Teacher's notes

Circle all the pyramids. Count the pyramids and write the number in the correct box at the top of the page. Then count the cones and write the number in the other box at the top of the page.

Date: _____

Sort 3D shapes

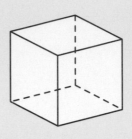

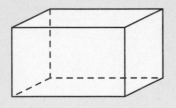

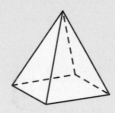

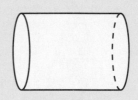

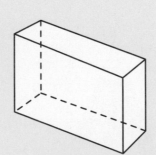

Teacher's notes

Colour **blue** all the shapes that have **6 faces**. Colour **red** all the shapes that have a **curved face/surface**.

Making shapes

Date: _____

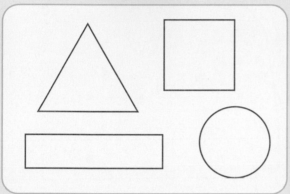

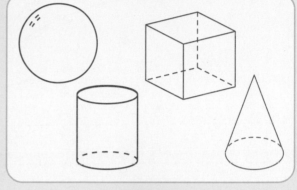

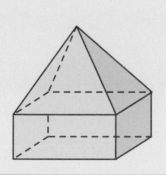

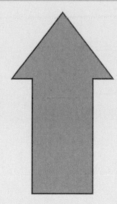

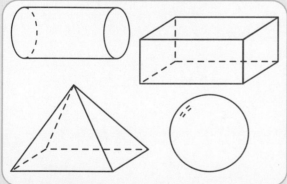

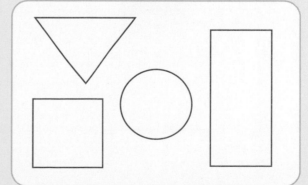

Date: _____

0 to 20

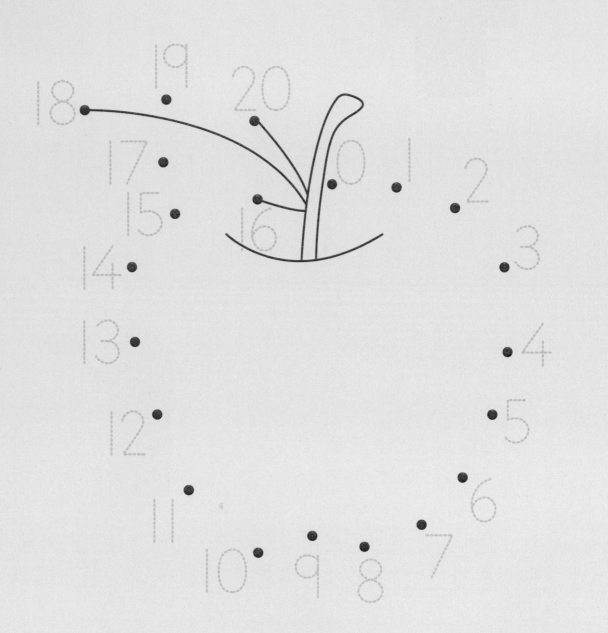

Teacher's notes

Trace each number. Then draw lines to join the dots in order, from 0 to 20.

20 to 0

Date: _____

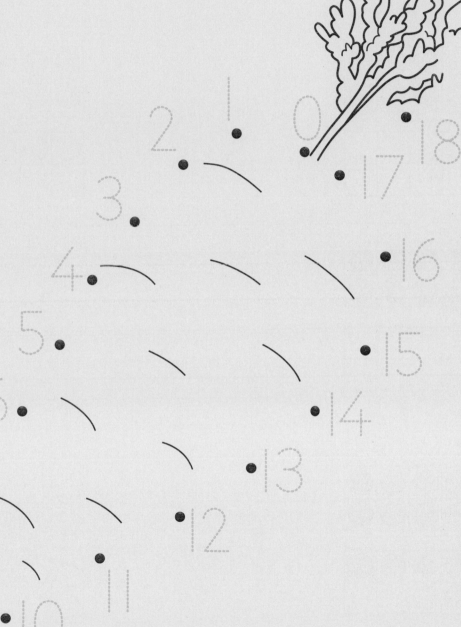

Teacher's notes

Trace each number. Then draw lines to join the dots in order, from 20 to 0.

25

Count and match

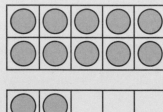

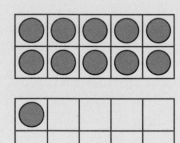

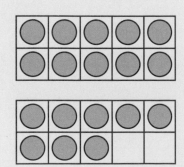

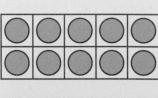

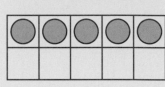

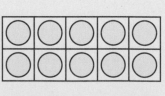

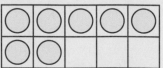

Teacher's notes

Count the dots on each pair of ten frames. Draw a line to the matching number. Then trace the number.

Fewer

Date: _____

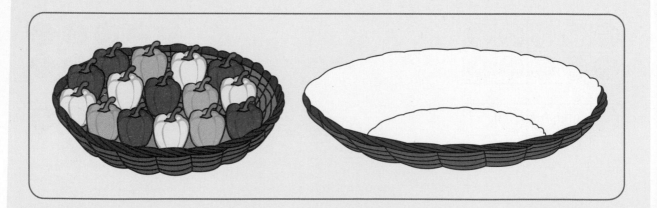

Teacher's notes

For each row, count the vegetables in the first basket. Draw **fewer** vegetables in the empty basket.

Date: _____

More

0	1	2	3	4	5	6	7	8	9	10
11	12	13	14	15	16	17	18	19	20

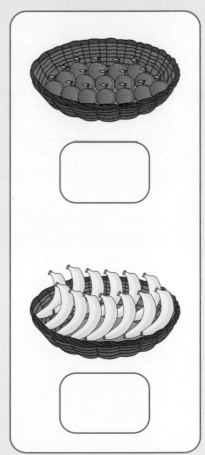

Numbers 0 to 10

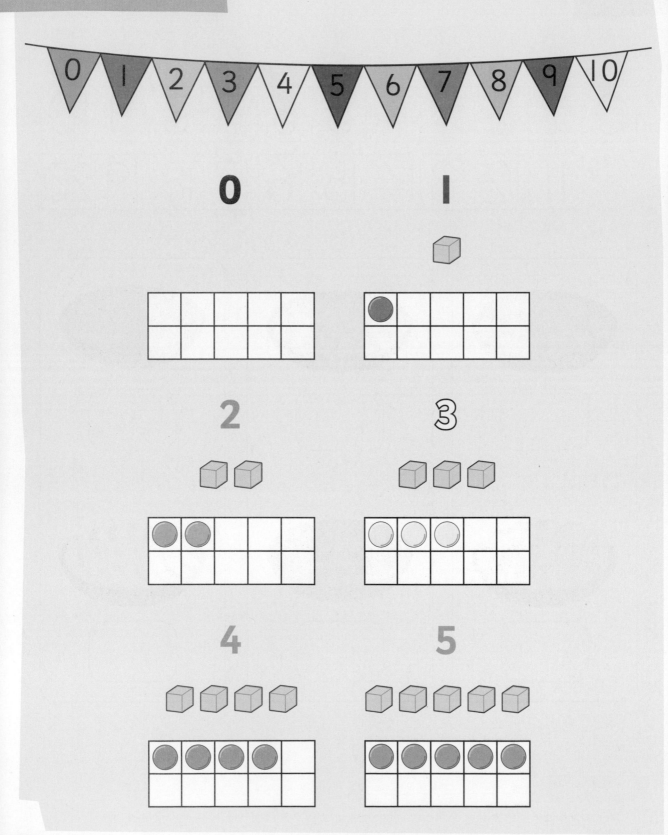

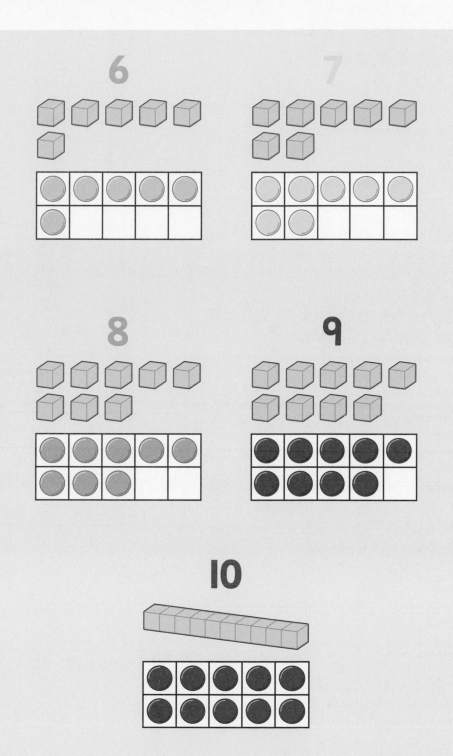

6

7

8

9

10

0	1	2	3	4	5	6	7	8	9	10

_____ has achieved these Maths Reception objectives:

Counting and understanding numbers

• Count on and back in ones, starting from any number from 0 to 10, then to 20.	1	2	3
• Count objects, actions and sounds from 0 to 10, then to 20.	1	2	3
• Recognise the number of objects presented in familiar patterns up to 10 without counting (subitise).	1	2	3
• Understand that zero represents none of something.	1	2	3
• Estimate a group of objects and check by counting.	1	2	3
• Understand and use ordinal numbers from 1st to 10th in different contexts.	1	2	3

Reading and writing numbers

• Read and write numbers from 0 to 10, then to 20.	1	2	3

Comparing and ordering numbers

• Understand the relative size of quantities to compare numbers from 0 to 10, then to 20.	1	2	3
• Understand the relative size of quantities to order numbers from 0 to 10.	1	2	3

Understanding shape

• Identify, describe, manipulate and sort 2D shapes, including reference to the number of sides and whether the sides are curved or straight.	1	2	3
• Identify, describe, manipulate and sort 3D shapes, including reference to the number of faces and whether faces are flat or curved.	1	2	3
• Compose and decompose shapes, recognising that a shape can have other shapes 'within' it.	1	2	3

Position, direction and movement

• Use everyday language to describe position, direction and movement.	1	2	3

Measurement

• Use everyday language to describe and compare length, height and width, including long, longer, longest, short, shorter, shortest, tall, taller, tallest, wide, wider, widest, narrow, narrower and narrowest.	1	2	3

Statistics

• Sort and match objects, pictures and children themselves, explaining the decisions made.	1	2	3
• Count how many objects share a particular property.	1	2	3

1: Emerging   2: Expected   3: Exceeding

Signed by teacher:
Signed by parent: Date: